THE FAST AND FEARLESS JOURNEY OF ESTEBAN OCON

Norman V. Parnell

Copyright @ 2024 By Norman V. Parnell.

All rights reserved. No part of this book may be reproduced, distributed, or transmitted in any form or by any means, including photocopying, recording, or other electronic or mechanical methods, without the prior written permission of the publisher, except in the case of brief quotations embodied in critical reviews and specific other noncommercial uses permitted by copyright law.

Table of contents

INTRODUCTION

A World of Speed

How Esteban Got Started in Racing

CHAPTER 1: THE BEGINNING OF A DREAM

French upbringing

First Steps in Racing

CHAPTER 2: RACING THROUGH THE YEARS

Karting Adventures

Getting Past Obstacles

CHAPTER 3: FORMULA 1: THE BIG LEAGUE

What is Formula 1?

Esteban's First F1 Race

CHAPTER 4: THE ROAD TO SUCCESS

Hard Work and Training

Key Races and Memorable Moments

CHAPTER 5: THE FEARLESS RACER

Bravery on the Field

Esteban's Team and Fans

CHAPTER 6: LIFE OUTSIDE THE TRACK

Hobbies and Interests

Family and Friends
CHAPTER 7: A CHAMPION'S LEGACY
Esteban's Impact on Racing
The Future of Racing
CONCLUSION

INTRODUCTION

A World of Speed

Imagine living in a world where the smell of tires burning on hot asphalt, the sound of motors roaring, and the wind blowing past you so fast that it feels like you're flying are all there. For many people, racing is more than a sport; it's a way of life, a passion, and a desire that outpaces the speed of the actual cars. Only the most courageous and accomplished drivers can handle the thrill, danger, and excitement of the track, where speed is everything.

It's simple to become enthralled with the cars' impressive speeds and vibrant designs as they pass by in a blur during a race. However, every race has a backstory of perseverance, hard work, and a passion for the game. Racing is more than just moving quickly; it's also about developing control, making snap judgments, and conquering obstacles that can alter drastically instantly.

There is no more incredible thrill than racing. Drivers experience an inexplicable high, a sense of empowerment and independence, and a sense of being the world's kings or queens. Every quick curve, every acceleration, and every victory is what makes racing so exciting. Millions of people are captivated by racing, whether it's the sound of the vehicle engine roaring before a race or the applause from the audience when the winner crosses the finish line.

One name that sticks out in this fast-paced world is Esteban Ocon. He is more than simply a racer; he is a part of that fantastic world and has experienced the highs and lows, the difficulties and victories, of being a Formula 1 driver. We must examine what initially pulled Esteban to this world—the excitement, the challenge, and the passion that racing requires—in order to comprehend his journey.

How Esteban Got Started in Racing

It wasn't always Esteban Ocon who drove Formula 1. His path to the racing world's top started modestly, like that of many great athletes. Esteban was a young lad with a lofty aim long before he was speeding around racetracks in some of the world's fastest automobiles. His ambition was to become the best, not merely to race. However, how can such a fantasy come to pass?

For Esteban, it all started when he was born in Échirolles, a little town in France. Esteban had a strong passion for racing even though he didn't have access to the expensive equipment or high-tech racing cars that many professional drivers might have when he was growing up. Despite their lack of wealth, his family had faith in him. Esteban's parents put a lot of effort into providing for him since they understood how unique his love of racing was.

His adventure began at the go-kart track, a modest, uncomplicated course where many well-known racers, including Esteban, start their careers. Go-karting was more than a fun pastime for Esteban; it was the beginning of something greater. Go-karts are small, quick, and need much expertise to maneuver. Despite his youth, Esteban had a remarkable innate ability to operate these little racing machines. Even as a young child, he stood out for his ability to drive the kart at fast speeds, accelerate precisely, and steer it through tight turns.

Esteban loved the challenge more than the racing itself. Each time he got in the go-kart, he gained fresh knowledge about pushing himself, overcoming challenges, and continuously improving. He saw racing as a jigsaw, with each race being a new puzzle piece to solve. What distinguished him was his desire to be the best and his insatiable curiosity.

However, Esteban's path was complex. Like all dreams, there were obstacles. Sometimes, he didn't win a race, things didn't go as planned, or it seemed impossible to continue along the road. However, Esteban persisted. He knew that improving daily and learning from failures were essential to being the best. He began to rise in the racing world due to his hard work and dedication.

Esteban was already causing a stir in competitive racing as a teenager. After joining the French racing squad, he quickly competed on far larger venues because of his innate talent and unwavering work ethic. Some influential figures in the racing industry took notice of his speed and abilities, and he soon received offers that most teenage drivers could only imagine. After competing in go-karts and cars, he eventually switched to Formula Renault, a highly competitive racing series that would put his abilities to the test.

He had an opportunity to prove himself in every event. And demonstrate that he did. He was fighting against the difficulties of being young, inexperienced, and an outsider in a world where many of the best racers had been competing for years, in addition to the clock and other drivers. However, Esteban's talent and desire were indisputable. Thanks to his triumphs in more minor racing leagues, he gained entry into the elite Formula 1 world, where only the most gifted and daring drivers can participate.

When Esteban joined Manor Racing's Formula 1 team in 2016, it was his big break. Although it wasn't easy, Esteban showed he belonged in his maiden season in the top racing league. By fighting bravely on the track, he demonstrated to the world that he was prepared for the demands of the highest level of competition. He stood out among the finest due to his innate talent, unwavering focus, and willpower.

He continued to make an impression in Formula 1 as his career developed, but joining Force India in 2017—now known as Racing Point—marked a significant turning point. There, competing against some of the world's top drivers, he really started to showcase his abilities. Although Esteban's path was not without its challenges, it was also full of amazing experiences, like his first podium finish in 2020, which he will always remember.

Esteban Ocon's journey is one of passion, tenacity, and an unrelenting pursuit of greatness in this world of speed and adrenaline. Esteban's journey from his early go-karting days to his ascent to the world stage in Formula 1 reminds him that dreams may come true—but only if you're prepared to put in the necessary effort. As we delve into Esteban's journey, we'll discover how his passion for racing, unwavering willpower, and bravery on and off the track propelled him to his current position.

CHAPTER 1: THE BEGINNING OF A DREAM

French upbringing

The narrative of Esteban Ocon starts on September 17, 1996, in Échirolles, a little village in the French Alps close to Grenoble. Although Esteban's family didn't have much growing up, they did have something far more significant: a strong love and belief in him. Despite having little racing experience, Esteban's parents, particularly his father, were aware of his early love of speed. His father did everything he could to foster Esteban's goals, even though he had no racing experience, while his mother worked constantly to provide for the family. They knew that Esteban's love of vehicles and speed was the beginning of something more significant.

From an early age, Esteban's world was centered around racing. Unlike other kids, he built race cars out of whatever materials he could find instead of using standard toys. With every year that went by, his passion for automobiles and their operation deepened. Esteban's attention became more narrowly concentrated as he grew older. He wanted to be behind the wheel and experience the rush of speed and competitiveness instead of merely playing with toy vehicles.

Even though many young boys aspire to be professional sports, not all are allowed to do so. Esteban's family wasn't wealthy, and the racing expenses might have been a significant barrier. However, Esteban's parents were adamant about helping him. Esteban's karting career was the first step toward his aspirations, and his father spent a lot of overtime to finance it. Although Esteban didn't begin in the most opulent circumstances, he had something more valuable: a commitment to self-improvement, a laser-like focus on the objective, and a steadfast faith in his abilities.

Esteban began karting at ten, which would be the basis for his racing career. He would accompany his father to nearby karting tracks to practice and compete. Most professional racers start their careers in karting, where Esteban would pick up the skills necessary to maneuver a car at fast speeds, negotiate turns, and compete against other drivers in a competitive setting. Esteban's innate aptitude was evident even in these early days. Everyone was impressed by his passion and perseverance as he swiftly rose through the ranks of local events.

He stood out from other children his age the first time he sat in a go-kart. He drove with a serene assurance and a profound awareness of how to challenge himself to improve with each race. Even though he was not an expert in racing, his father supported and encouraged him during these early years. Esteban excelled in karting not just because of his natural talent but also because of his strong work ethic, ability to concentrate, and unwavering willpower.

Esteban's sense of competition grew stronger as he got older. By the time he was in his teens, he was winning races against some of France's top young kart racers and racing locally. Every triumph made him more determined to compete in Formula 1 one day. But there were still a lot of challenges to overcome. Esteban had a difficult journey to becoming a professional racer. He had to show himself as a rookie driver in every race against other drivers and the sport's financial challenges. His family's sacrifices, along with his innate talent and motivation, enabled him to persevere despite the high expenses of racing.

His upbringing in Échirolles gave him a distinct viewpoint on racing. Despite being surrounded by breathtaking scenery and tucked away in the heart of the French Alps, the town lacked the resources and connections other would-be racers could have had in larger cities. Esteban's tale is one of perseverance—turning that passion into a reality by seizing every chance, no matter how tiny. His modest childhood in Échirolles was the foundation for all his future accomplishments.

Esteban's quest was about more than simply racing; it was about demonstrating that hard effort, commitment, and a strong sense of self-worth could make dreams come true. During his formative years, which took him from a rural hamlet to the bright lights of Formula 1, Esteban Ocon learned a lot about tenacity, commitment, and the value of family support.

First Steps in Racing

Although Esteban's initial forays into racing were not particularly glitzy, they were characterized by the tenacity that would represent the rest of his career. He didn't have the riches many of the best racers had when he was younger, but a desire and an unquenchable drive drove him. At ten, when many aspiring Formula 1 drivers were already well-established in the karting scene, he started his career in competitive racing.

Esteban drove a go-kart on a small neighborhood track close to his hometown for the first time. This would have been a delightful weekend pastime for many kids, but it marked the beginning of something much more significant for Esteban. Go-karting was more than simply a pastime to him; it was his pass to something bigger. Despite having few resources initially, his love of racing compensated for them. Even though he had no racing experience, Esteban's father remained his staunchest ally, accompanying him to several courses and doing everything they could to realize his ambition.

His natural talent was evident in those early days. He could already operate the kart with exceptional precision while many other young drivers still figured out their rhythm. Karting demands extraordinary focus, balance, track reading skills, and speed. From the beginning, Esteban had all of these abilities. He could speed out of curves like he had been racing for years, handle tight turns gracefully, and maintain focus throughout each lap.

Esteban began participating in more significant regional competitions, where the level of competition was significantly higher, as he rose through the karting levels. Some of France's top young drivers attended these races, and Esteban immediately established himself. Despite being one of the track's youngest drivers, Esteban showed promise by competing against older, more seasoned drivers. He wasn't merely taking part; he was competing against the best drivers and frequently outperforming them.

However, mastering the ins and outs of the sport is just as important to racing as having driving talent. Esteban discovered via karting how crucial it was to control not just his vehicle but also his strategy and mental fortitude. High-speed racing requires fantastic concentration, and Esteban quickly discovered that winning each race required more than just driving skill—it also required having the correct attitude. He needed to develop his ability to manage stress, remain composed when things didn't go as planned, and recover from errors.

Esteban's innate skill and tenacity in karting were not overlooked. A year after he began, in 2011, he advanced to a more competitive level by competing in the French Formula 4 Championship, a significant milestone for many young drivers. Esteban had the opportunity to demonstrate his ability to handle both full-sized race cars and go-karts in Formula 4. It was difficult for him to adjust to the world of auto racing, but Esteban was capable. He could modify his racing style to fit the more significant, quicker vehicles since he had the necessary expertise and talents.

He had fierce competition in his first season of Formula 4, but his maturity and focus won over everyone. Even though go-karting and auto racing are different, Esteban demonstrated that his karting background had equipped him with the abilities he needed to be successful. Teams searching for the next great driver noticed him because of his repeated good race results. Although the victories weren't easy, Esteban stood out due to his reliable performances and ability to remain composed under duress.

Esteban discovered a crucial lesson in his early racing days: winning races isn't the only way to succeed. It has to do with development. Every race served as a springboard, allowing him to hone his skills, get more experience, and prepare for the difficulties ahead. Esteban started to see the road to professional racing become more evident in Formula 4. His progress toward his ultimate objective—Formula 1—would be significantly aided by the little triumphs, the knowledge gained from setbacks, and the relationships he formed.

CHAPTER 2: RACING THROUGH THE YEARS

Karting Adventures

Esteban Ocon's racing career began with his karting exploits, which also helped to mold him into the driver he would become. As soon as he sat behind the wheel of his first go-kart, Esteban became engrossed in the world of speed, accuracy, and competition. For him, it was more than a sport; it was a chance to succeed, develop, and learn.

When Esteban was about ten years old, he started karting seriously. Unlike many of his rivals, Esteban didn't come from a wealthy family that could afford to buy him the most costly gear or send him to the top racing schools. He began with a basic kart, but his unwavering commitment and innate talent more than made up for his lack of money. He spent hours honing his abilities and learning the sport's nuances at nearby karting courses.

Karting is about learning the skill of control, not just about racing. Esteban concentrated on understanding how a go-kart handled each turn, how to go around sharp turns, and how to apply the throttle smoothly during those early races. He had to maintain his composure throughout each lap, no matter how exhausting or difficult the race got, making it a test of mental and physical stamina.

Esteban advanced swiftly through the local karting competition ranks, competing against other young drivers who were as passionate about racing as he was. During these initial days, he faced fierce competition. However, Esteban had a special capacity to concentrate on his objectives under increasing strain. His ability to remain composed on the track became one of his defining characteristics, enabling him to perform effectively continuously in spite of obstacles.

When Esteban started competing in national karting events, it was one of his racing career's most significant turning points. There was much more competition, with some of France's brightest young drivers vying for the top positions. Esteban, however, was unfazed. He developed his ability to manage the mental and physical strains of karting with each race. He had to use strategic thinking to manage tire wear, position himself for the best racing lines, and determine when to push and save energy. He would carry these lessons with him throughout his professional life.

His innate talent started to show during his karting career. After winning multiple regional and national karting championships, he became one of the sport's emerging stars. However, Esteban never took his accomplishments for granted. He persisted in pushing himself further, constantly seeking methods to improve, whether honing his technique or researching the strategies of other elite drivers to see how they handled competitions.

When Esteban participated in the renowned CIK-FIA European Racing Championship in 2012, it was a significant turning point in his racing career. The top karting drivers competed in this event, and Esteban's performance was outstanding. Several elite racing teams were interested in him after he competed against some of the best karts of his age and placed first. Esteban was more than just a local prodigy; he was a driver with enormous potential who would one day compete at the top motorsport levels.

The relationships Esteban formed along the way were as meaningful to karting as the actual races. He began to meet other drivers, trainers, and mentors as he became well-known in the karting community; these individuals would later be crucial in his transfer to car racing. As he advanced through the levels of competitive racing, the connections and friendships he formed during these formative years proved pivotal.

Through his karting experiences, Esteban cultivated the competitive mindset and technical abilities necessary to succeed as a racer. He gained skills in managing the highs and lows of racing, overcoming obstacles, and approaching every task with composure and concentration. These experiences gave Esteban a strong foundation that would help him in the larger and more intricate world of auto racing.

Although they were characterized by fierce competition, long practice sessions, and challenging races, Esteban Ocon's karting years were also pivotal in helping him define his career. He significantly impacted the karting world by demonstrating that a young driver could achieve great things with perseverance, hard effort, and a passion for the sport. One kart race at a time, Esteban was already setting the foundation for his future triumphs, even though the world of Formula 1 may have looked far away.

Getting Past Obstacles

Like many elite sportsmen, Esteban Ocon overcame several challenges before becoming a Formula 1 driver. Many challenging times put his mental and physical fortitude to the test along his journey to the top motorsport levels. Esteban's narrative, which includes financial hardships, intense competition, and setbacks, is as much about overcoming adversity as it is about outpacing others.

The cost of pursuing a racing career was one of the biggest obstacles Esteban had to overcome initially. Racing is a costly sport, particularly at the top levels. The expenses of go-karting, advancing to more competitive racing levels, and keeping up with the required gear were prohibitive. Esteban's family had to make significant sacrifices to support his dream because they weren't wealthy. They were sometimes still determining how they would support his racing career. However, despite these financial difficulties, Esteban's parents never gave up on him. While his father took on several jobs to help pay the bills, Esteban was focused on giving every race his all and refused to let money stop him. His increasing ability and sense of determination would be crucial in helping him overcome the obstacles.

His family's sacrifices drove Esteban, but his next obstacle was the intense rivalry he encountered while advancing through the ranks. As he advanced to Formula 4 and started competing in increasingly competitive karting competitions, he encountered exceptionally gifted drivers with more excellent money and support than he did. These racers had access to bigger teams, more excellent gear, and more sophisticated training. Esteban, however, was forced to rely on his hard work and innate talent. Every race was, therefore, a test of his resolve and ability to maintain concentration in the face of overwhelming odds, in addition to his driving prowess.

Esteban had to deal with the internal struggle of maintaining motivation and focus in addition to the external constraints. It needs mental toughness, which many young drivers lack, and physical conditioning to compete at such a high level. There were moments when he questioned whether he could go on because of the strain of representing his family and nation. But his passion for the game and his conviction that every obstacle was a chance to get better gave him comfort. Esteban's ability to bounce back from setbacks and use them as teaching moments rather than allowing them to keep him back was a typical example of his resilience.

His ascent through the elite single-seater racing ranks was one of the most significant obstacles in his early career. After dominating karting, he had to establish himself in a different setting, so he raced in junior single-seater championships like Formula Renault and Formula 3. Despite his talent, Esteban needed help to adjust to the more sophisticated equipment and fiercer competition. Because these vehicles were faster, they required different skills, such as coordinating strategy with his team, handling tire deterioration, and handling longer races. The pressure to do well for himself and the teams and sponsors supporting him made every race even more intense.

However, Esteban demonstrated his abilities in these junior divisions, rapidly picking up the skills necessary to handle the challenges of single-seater racing. Although he had a difficult transition from karting to car racing, he advanced quickly through the ranks thanks to his strong work ethic and unwavering drive for the better. He gained knowledge and confidence, enabling him to take on increasingly complex tasks with every victory, podium finish, and lap.

When Esteban switched to Formula 1, he had yet another significant challenge. The challenge of racing in the most prestigious motorsport series in the world was enormous, even after securing a berth with Manor Racing in 2016. Esteban had to learn how to get the most out of Formula 1 cars because they were more advanced and powerful than anything he had ever driven. In addition to adjusting to the difficulties of driving a Formula 1 car, his first season was a learning experience as he had to master the intricate dynamics of working with a vast team, coordinating with engineers, and optimizing the car's performance for every race. Esteban had to make every second matter, whether in practice, qualifying, or the actual races because he didn't have the luxury of being on a top team.

Despite the difficulties, Esteban's tenacity and flexibility were evident. Throughout his Formula 1 career, his determination to overcome hardship was a recurring theme. He moved to Force India (now Aston Martin) after his time with Manor, where he had to contend with some of the top drivers in the sport, including Sergio Pérez, a colleague. Esteban had to show himself against an experienced driver during this pivotal time, and he also had to learn how to maximize the car's performance—a critical ability for any Formula One driver. Esteban's position in Formula 1 was cemented by his innate talent, increasing experience, and mental toughness.

Despite these difficulties, Esteban never lost sight of his passion for racing. He never let losses define him; instead, he used them as opportunities for personal development. His tale is an encouragement to everyone who encounters hardship in their endeavors and a monument to the strength of tenacity. A significant part of Esteban's growth as a driver and a major contributor to his success on the international scene was his ability to overcome obstacles, whether they be monetary, competitive, or mental.

CHAPTER 3: FORMULA 1: THE BIG LEAGUE

What is Formula 1?

Formula 1 (F1) is the ultimate single-seater, open-wheel motorsport, one of the world's most prestigious and technologically sophisticated racing series. An international tournament known as the **Formula 1 World Championship** is held yearly and consists of a series of races. Top-tier drivers and automakers compete worldwide, and the teams and vehicles push the speed, accuracy, and technology boundaries.

Fundamentally, Formula 1 is a measure of engineering prowess and driver skill. With a focus on speed, aerodynamics, and precise handling, Formula One cars are made especially for high-speed racing. These are some of the most sophisticated devices on the planet, driven by hybrid power units that optimize performance and minimize environmental impact by combining electric motors and conventional combustion engines. These vehicles have a top speed of 200 mph and can accelerate from 0 to 60 mph in seconds. The racetracks are diverse and might include purpose-built courses like Silverstone or Suzuka and city circuits that meander through the streets of well-known towns like Singapore and Monaco.

Many **Grand Prix** (GP) races on the Formula 1 calendar are contested in various nations worldwide. The races are contested on temporary tracks, public roads, and permanent circuits during the season, usually lasting from March to December. The **Drivers' and Constructors' Championships** are the ultimate goals of the Formula One season, and while each Grand Prix is a stand-alone race, points are given depending on finishing positions. At the end of the season, the driver with the most points wins the Drivers' Championship, while the team (two drivers) with the most points overall wins the Constructors' Championship.

Multiple laps around the track are a feature of Formula 1 races, which are usually intense and last 1.5 to 2 hours. Pit stops, where teams replace tires and make tweaks to the cars, quick speeds, tight turns, and clever tire management are all features of the actual races. Teams of highly qualified engineers, technicians, and strategists collaborate to offer their drivers the best opportunity for success. Pit stops are rapid, sometimes taking less than two seconds, demonstrating the degree of accuracy and coordination needed for the sport.

Formula 1 incorporates intricate strategic factors in addition to the technical components of the cars. Weather, track surface, and tire wear are just a few variables that teams must continuously adjust to. Drivers must manage fuel, tire wear, and race strategy while making snap decisions that balance the necessity for speed and the dangers of exerting too much pressure.

With renowned drivers like Michael Schumacher, Ayrton Senna, and Lewis Hamilton, as well as classic teams like Ferrari, Mercedes, Red Bull Racing, and McLaren,

Formula 1 boasts a rich history. Over the years, the sport has witnessed remarkable technological advancements, ranging from the advent of turbocharged engines to state-of-the-art hybrid power units and sophisticated aerodynamics.

The intense competitiveness is just as distinctive as the speed and technology in Formula 1. The sport combines cutting-edge innovation, teamwork, and individual talent. Millions of fans worldwide watch the races live or on television, making it a global spectacle. By uniting fans, drivers, engineers, and manufacturers worldwide through their shared love of speed, accuracy, and quality, Formula 1 goes beyond the confines of motorsport.

Esteban's First F1 Race

A turning point in Esteban Ocon's career, his entry into Formula 1 was evidence of his progression from karting to the top motorsports league. In 2016, he competed in his maiden race at the legendary **Circuit de Spa-Francorchamps** for the **Belgian Grand Prix**. In addition to being Esteban's first appearance in Formula 1, it was also a momentous occasion because he was joining **Manor Racing**, a team going through a transformation. Even if Manor wasn't regarded as one of the best teams, Esteban had a valuable chance to showcase his skills on a global scale.

Esteban's maiden Formula One race had been heavily anticipated. He had developed his abilities over several years in junior racing classes, such as Formula 3 and Formula Renault, where he gained a reputation as a fast and reliable driver. In 2015, he also worked as a reserve driver for Mercedes, where he learned more about Formula One equipment and race weekends. Nothing, however, could adequately prepare him for the extreme speed and intricacy of an F1 car during a race weekend.

He was taking over for **Rio Haryanto** at Manor Racing for the second half of the 2016 season, going into the Belgian Grand Prix. Compared to the best teams like

Mercedes, Red Bull, or Ferrari, the team needed more funding, and the vehicle, the **Manor MRT05**, was required to be competitive. Esteban knew this was his chance to prove himself but was determined to take advantage of it.

A novice might find the Belgian Grand Prix Spa-Francorchamps intimidating because of its difficult layout, erratic weather patterns, and fast-paced parts. The course is well-known for its long straightaways that challenge driver skill and vehicle performance, dramatic elevation changes, and sweeping turns like Eau Rouge. Even though the Manor car was not anticipated to contend for the top spots, Esteban saw it as an opportunity to showcase his potential and adaptability.

Esteban's performance on race day was outstanding. Although it wasn't great, he qualified in **19th place**, which was a respectable finish given the Manor car's restrictions. Despite the uproar of his debut Formula 1 race, he maintained his signature poise throughout the race. Esteban encountered difficult circumstances during the race, including a spectacular **rain shower** that made the track slick. He demonstrated a fantastic ability to control his speed, avoid errors, and maintain concentration under duress throughout the race. His performance demonstrated his talent even though he did not finish in the points. Esteban made a solid start for the young Frenchman, finishing in 16th place, ahead of his colleague **Pascal Wehrlein**.

The most remarkable aspect of Esteban's debut race was his ability to keep up with more seasoned competitors. He frequently displayed patience and made wise choices while racing. He displayed the poise that would come to define his career and a level of maturity beyond his years. He was already accustomed to the demands of F1 teams, engineers, and the race weekend schedule from his time as a reserve driver with Mercedes, which helped him adjust to the event well.

Esteban's first race was a turning point in his career, even if it did not yield any points or a noteworthy outcome. It demonstrated that his skill set was prepared for the demands of the sport's top level and that he could manage the tremendous pressure of Formula One racing. More significantly, his first race helped establish him as a driver to watch, drawing interest from both fans and Formula One teams.

More than just a debut, Esteban's first Formula 1 race marked the start of an incredible career. Since then, he has kept developing his skills and reputation, eventually landing a spot with **Force India** (now Aston Martin) for the upcoming seasons and going on to have great success in the sport. Even if his starting position in Belgium wasn't noteworthy, it was an essential beginning to a lengthy and fruitful Formula 1 career.

CHAPTER 4: THE ROAD TO SUCCESS

Hard Work and Training

Esteban Ocon's ascent in Formula 1 has been attributed mainly to his diligence and preparation. Esteban has continuously shown that his success isn't solely down to skill; it also stems from a tireless work ethic and a dedication to getting better, as evidenced by his early karting and stint in Formula One. Professional racing drivers lead complex lives that need more than just physical prowess. It involves mental acuity, concentration, and the capacity to adjust to novel situations continuously. One of the most critical aspects of Esteban's career has been his dedication to training, both on and off the track.

Esteban's physical fitness is one of the most critical components of his training. Formula One drivers must endure severe G-forces, fast cornering, and extended periods of intense concentration during races. Esteban has created a thorough exercise program to meet these needs. His training regimen includes strength training for neck and core stability, cardiovascular exercises to increase endurance, and reaction drills to improve reflexes. An F1 driver's neck muscles are essential because they must endure the forces used when cornering. To maintain control of the car during the race, Esteban devotes much time to exercises that strengthen his upper body and neck.

In Formula 1, mental fitness is just as necessary as physical preparation, and Esteban has always prioritized this area of his training. High-level racing demands extraordinary focus and the capacity to remain composed under duress. Esteban has improved his resilience and attention by working with mental coaches and sports psychologists. Esteban can prepare for the psychological demands of racing through these sessions, which frequently include mental drills, visualization techniques, and race scenario simulations. His mental agility enables him to make snap judgments when faced with unforeseen race conditions or high-pressure circumstances like overtakes.

Esteban is renowned for his profound comprehension of the technical aspects of racing, his physical prowess, and mental preparation. Because Formula 1 cars are intricate machinery, drivers frequently have a say in tire strategy, race tactics, and car setup decisions. Esteban has worked with his engineers for hours to grasp the nuances of automobile performance and mechanics. He is skilled at communicating with his staff to ensure his car is configured to operate at its best. Because of his technical expertise, he can provide his engineers with insightful criticism that helps them improve the car during practice or between races.

His dedication to diligence is also shown in the amount of time he spends using simulators. Teams use very sophisticated F1 simulators to simulate actual race conditions. Esteban spends endless hours using these simulations to improve his driving abilities, learn new tracks, and research race tactics. In the off-season, simulators also help him maintain his best mental state and ensure he's constantly prepared for the next race, no matter what obstacles or circumstances may come up.

The actual race weekends demand excellent planning, and Esteban is renowned for his meticulous methods. He thoroughly examines the track before each race, looking at historical performance information and his team's tactics. Throughout the practice sessions, Esteban concentrates on getting used to the track and adjusting the car's setup to the circumstances. His readiness and diligence are evident in his ability to swiftly grasp the subtleties of a circuit and convey this knowledge to his colleagues.

Rest and recuperation are also a part of Esteban's training schedule. Racing may be mentally and physically taxing; recovery is as crucial as the actual work. Esteban adheres to a strict recuperation regimen that includes stretching exercises, physiotherapy, and sleep to avoid injuries and maintain his body in optimal shape. Esteban maintains top form for every race and prevents burnout during the lengthy Formula One season, thanks to this balance between hard work and healing.

Esteban Ocon has made a name for himself as a disciplined and well-rounded driver through his commitment to hard work and ongoing training. His success in Formula 1 reflects his unwavering drive for greatness in every facet of his career, not just his innate talent. Esteban's dedication to progress has been essential to his success in motorsport, whether in terms of race strategy, technical understanding, mental preparation, or physical conditioning.

Key Races and Memorable Moments

Throughout his Formula 1 career, Esteban Ocon has had several significant races and noteworthy events, demonstrating his talent, tenacity, and resolve. In addition to showcasing his brilliance, these instances show that he can perform under pressure and compete against some of the world's top drivers. Let's examine a few of these noteworthy points in Esteban's trip.

During the **Sahara Force India team's surprise renaming to Racing Point** In 2018, Esteban experienced one of his most unforgettable moments. Esteban and his teammate, **Sergio Pérez**, engaged in an exciting duel at the **Bahrain Grand Prix**. Esteban had a chance to demonstrate his abilities against a more seasoned teammate, making it a pivotal point in his career. Esteban finished in **8th place** after putting on a solid showing both during the weekend and the actual race. In addition to earning him essential points, this outcome enhanced his reputation as a competent and competitive driver.

The **2019 German Grand Prix at Hockenheim**, a race notorious for its mayhem and rainy circumstances, was another crucial event for Esteban. At this moment,

Esteban was representing **Renault**, and the race became one of the most unpredictable in recent memory as several drivers made mistakes and spun out. Esteban maintained his poise and skillful driving amid the turmoil, finishing in **7th place**. This outcome was significant because it showed that he could handle challenging situations and score points in trying conditions—a skill that Formula 1 highly values. In a circumstance when many others failed, it also brought him respect for his maturity and racecraft.

Esteban's career was further defined during the **2020 Sakhir Grand Prix**. After battling for years to produce consistent results in mid-tier teams, Esteban finally had his moment to shine. Esteban, racing for **Renault** with Daniel Ricciardo, had to contend with a very competitive field in an exciting race. The race featured multiple lead changes, dramatic collisions, a red flag, and a dramatic pit strategy. Throughout the race, Esteban demonstrated his tenacity and intelligent thinking. Despite several obstacles, Esteban made the proper decisions at crucial times to maintain his position in the race. His first-ever podium finish in Formula 1 came in **2nd place**. This outcome signaled a significant turning point in Esteban's career and showed he had the skill and will to battle at the front of the grid.

The **2021 Hungarian Grand Prix**, where Esteban won his first-ever **Formula 1 victory**, was one race that stuck out for him. Although Esteban had been improving while racing for **Alpine**, few predicted he would win because of the supremacy of teams like Red Bull and Mercedes. Esteban was in a good position following a wild first lap where several cars collided. He maintained his lead throughout the race by making calculated choices and seizing openings. Esteban held off the competition and won the race with flawless tire management and a strategically placed pit stop. In addition to being a turning point in his career, his maiden victory demonstrated his endurance, racecraft, and capacity to seize chances when they present themselves.

Esteban has been crucial to the team's success and achievements. His triumph was important for Alpine (previously Renault) as a team at the 2021 Hungarian Grand Prix. The French squad achieved a historic victory, and Esteban's ability to remain composed under duress was essential to their achievement. This triumph demonstrated that Esteban could win in Formula 1 under the correct circumstances and resulted from years of perseverance and hard work.

Both fans and other drivers respect Esteban for his perseverance and passion. In the competitive world of Formula 1, he stands out as a driver due to his ability to execute in erratic conditions, his poise under pressure, and his steady growth over the years. Future generations of drivers will probably continue to be inspired by these critical races and unforgettable situations.

CHAPTER 5: THE FEARLESS RACER

Bravery on the Field

Esteban Ocon's incredible bravery on the track, which he has often shown throughout his racing career, is what defines his Formula 1 career. He has gained the admiration of both supporters and rivals for his capacity to persevere through trying times, make audacious choices, and face obstacles head-on. In addition to his physical boldness, Esteban possesses mental tenacity and the desire to take chances when the chance presents itself. Whether in a high-speed pursuit or under extreme strain, his ability to bounce back from setbacks has been a career hallmark.

His bravery during the **2020 British Grand Prix** at Silverstone was one of the most notable instances. Esteban faced some of the world's top drivers in the race, and the fast-paced track was a formidable obstacle. Rain threatened to stop the race at one stage, forcing Esteban to cope with the leading competitors and the weather. His fearlessness was clearly displayed when he made daring, calculated efforts to defend his position—often risking everything to control the car. In addition to helping him earn essential points, his daring attitude to racing in challenging circumstances demonstrated his increasing self-assurance in his capacity to compete at the top level.

Esteban's **overtaking maneuver in the 2021 Emilia Romagna Grand Prix** is another instance of his bravery. During an intense rivalry season, Esteban made several bold maneuvers to challenge and pass several seasoned drivers. His forceful but controlled overtakes demonstrated his willingness to take measured chances when needed. When Esteban drove his car to the limit to get points for **Alpine**, this aggressive strategy paid off, and his team made notable progress in the constructors' rankings. He showed he could compete with the best in the sport while remaining composed and confident by taking on competitors like **Lewis Hamilton** and **Max Verstappen**.

Esteban's bravery is not limited to the racetrack. Another indication of his mental toughness is his capacity to overcome hardship, such as the challenging seasons spent with lower-tier teams. After leaving **Force India** In 2018, Esteban had to establish himself again during uncertainty. He didn't lose confidence in his skills even if he didn't have a seat for the 2019 season. Instead, he put in a lot of overtime as a reserve driver for **Mercedes** to prepare for more possibilities. When **Renault** signed him for the 2020 season, his perseverance paid off, and he displayed great bravery in establishing his value to the squad.

His first victory at the **2021 Hungarian Grand Prix** demonstrates Esteban's bravery in grasping opportunities. During the race, he was under constant pressure from more seasoned drivers and collaborated with his team to make snap decisions that would determine the result. Esteban had to deliberately manage the race and repel challenges from several drivers to win, so it was a challenging victory. His first victory was primarily attributed to his ability to concentrate and make brave choices under pressure.

In times of intense competition, Esteban's bravery is also evident. Esteban battled valiantly for positions in the 2021 **Austrian Grand Prix**, never giving up a challenge even though he was up against faster cars. Esteban's courage on the track enabled him to take full advantage of any circumstance, even when driving a less competitive car than some of his competitors. He got the most out of his automobile and earned essential points for **Alpine** because of his strategic thinking and desire to fiercely defend his position.

His fortitude is demonstrated on the racetrack, as is his unwavering dedication to becoming a better driver. He has consistently strived to improve his abilities and adjust to shifting racing circumstances. Thanks to his bravery in stepping outside of his comfort zone, he has developed from a young, bright prospect to an experienced Formula 1 driver who can contend for podiums and victories.

Esteban Ocon's courage on the track is characterized by his readiness to face obstacles head-on and follow his gut when it counts most rather than by flashy maneuvers or careless choices. Esteban has repeatedly shown that his heart and head are just as strong as his driving prowess, whether in the battle for a midfield spot, taking chances on the racetrack, or recovering from challenging times.

Esteban's Team and Fans

In addition to his driving prowess, Esteban Ocon's journey through Formula 1 has been influenced by his close bonds with his teams and supporters. In many respects, his success on the track is a result of the cooperation and support he gets from his racing teams and the devoted fan base that supports him.

Working closely with his crew was something Esteban rapidly learned from his early days in **Manor Racing**. To compete, Manor, a team not recognized for having a large budget, frequently had to rely on creativity and resourcefulness. Esteban's growth as a driver was greatly aided by his capacity to relate to the engineers and mechanics and provide constructive criticism while maintaining a cheerful disposition. Despite the team's financial constraints, Esteban's professionalism and teamwork helped lay the groundwork for his future in the sport. Even though he didn't win any games or take home any podiums, his time with Manor was vital in helping him understand the value of teamwork and communication.

Esteban found himself in a far more competitive environment after relocating to **Force India** (later rebranded as **Racing Point** and subsequently **Aston Martin**). Frequently struggling for points against teams like **Renault** and **McLaren**, the team was engaged in midfield combat. His relationship with colleague **Sergio**

Pérez was intriguing; the two drivers always encouraged one another to improve. Esteban had to swiftly adjust to the team's culture, which strongly emphasized productivity and outcomes. Esteban gained a competitive edge in this setting. He earned a reputation for getting the most out of the car, frequently going above and beyond expectations even though he didn't always have the fastest equipment.

The next significant step in Esteban's career was joining **Renault** in 2020, which signaled the start of a new phase. Esteban had the chance to collaborate closely with Renault, a team striving to regain its position at the sport's top. There was mutual respect between Esteban and his engineers at **Renault**. He established good chemistry with the crew almost away, and his insightful criticism improved the car's performance and setup. His flexibility in responding to the team's changing tactics and vehicle advancement was crucial to the team's overall success throughout the season. Even in the face of intense competition from other midfield teams,

Esteban's professionalism and work ethic were crucial to the team's success.

When **Alpine** (the renamed Renault team) started to demonstrate greater competitiveness in 2021, Esteban's collaboration with **Fernando Alonso** emerged as one of the season's most intriguing elements. Esteban got knowledge from one of the best drivers in the sport because of Alonso's extensive expertise and championship-winning background. Despite their rivalry, they had a respectful relationship in which both drivers encouraged one another to improve. Esteban's maturity and adaptability were evident in his ability to collaborate with such a talented partner. In addition to learning from Alonso, he became more self-assured after realizing he could compete successfully against one of the greatest in the sport.

In addition to his teams, Esteban's supporters have been crucial to his success. Esteban has a devoted and ardent following despite having a more subdued and introverted attitude than some of his peers. Many have been inspired by his journey through Formula One, from being a young, talented karter to achieving his first podium and, eventually, his first victory. His fans value Esteban's work ethic, tenacity, and fortitude in conquering obstacles. His fans, who had been rooting for him through years of arduous effort, disappointments, and near-misses, found his triumph at the **2021 Hungarian Grand Prix** very meaningful. That victory felt like a well-earned and long-awaited prize to Esteban's supporters.

Esteban's relationship with his fans has become increasingly reliant on social media. He shares moments from his professional life, training regimens, and personal life on social media sites like Instagram and Twitter. Esteban's posts frequently highlight his humorous nature and express his appreciation for his supporters. He is popular among fans because his sincere approachability and interactions with them deepen the connection between driver and fan.

Beyond his career achievements, Esteban has a strong bond with his team and supporters. He is renowned for his modest demeanor and gratitude to those who support his success. Esteban's genuineness is evident whether he shares a memorable moment with his supporters or praises his engineers following a fiercely contested race. One of the reasons he is so prevalent in the paddock and among the general public is his ability to balance his professional emphasis and personal relationships.

Esteban, his teams, and his supporters mutually connect in many respects. His skill and diligence fuel his team's success, but his supporters' encouragement and love inspire him to keep going. Esteban's growth has been greatly aided by the respect he shares with the teams he has worked with, and his fans' admiration for him keeps him motivated to keep improving in his Formula 1 career.

CHAPTER 6: LIFE OUTSIDE THE TRACK

Hobbies and Interests

Esteban Ocon maintains a healthy balance by engaging in various activities and pastimes outside his demanding racing profession. Although his main emphasis is motorsport, Esteban has frequently discussed the value of relaxation and how he spends his leisure time rejuvenating and following his passions.

Being physically fit and active is one of Esteban's primary passions. His dedication to physical training permeates various facets of his life and his preparation for Formula 1. In addition to his typical F1-specific exercises, Esteban likes trekking and cycling. In addition to keeping him in top physical shape, these pursuits allow him to unwind and detach from the demands of competition. He particularly enjoys cycling since it keeps him in shape while allowing him to experience nature and varied locations. In contrast to the fast-paced world of Formula 1, cycling provides Esteban with a different kind of thrill, whether on a mountain route or along coastal roads.

Traveling is one of Esteban's other interests, and it fits well with his F1 way of life. Because he travels to different places for racing, Esteban likes to experience the diversity of each Grand Prix location, try new foods, and learn about new cultures. Whether it's through images or anecdotes of the various places he's been to, he frequently shares his travel adventures with his admirers. For Esteban, travel is about more than just the races; it's about seizing the chance to see the world in a way many others would not be able to.

Esteban has talked about his passion for video games and is also a huge gamer. Playing video games is a chance for him to decompress and have fun after a demanding training session or racing weekend. He enjoys playing competitive online games and has even shared his gaming experiences with fans and other drivers, demonstrating that his interest in gaming goes beyond straightforward enjoyment. Additionally, playing video games helps him improve his hand-eye coordination and reaction times, two abilities that will aid him in his racing career.

Apart from these pursuits, Esteban's love of cars extends beyond racing. Although Formula 1 is the focus of his career, Esteban has stated that he loves automobiles of all types, particularly ones with performance potential. He likes to go to auto shows and is especially fond of supercars, which he frequently posts about on social media. Esteban views cars as racing tools and a profoundly valued form of engineering and design.

Additionally, Esteban is really interested in charity work. Despite his career as a racer, he frequently participates in humanitarian endeavors. He has backed issues pertaining to environmental preservation, children's health, and education. Thanks to his experiences and platform as an F1 driver, he has the chance to raise awareness of significant issues and make a constructive contribution to the community. His dedication to giving back wherever he can reflects his belief in using his celebrity for good.

More personally, Esteban likes to hang out with his friends and family. Despite his busy race schedule and international travel, family is still very important to him. Throughout his career, Esteban's strong relationship with his family has been a pillar of strength, and he regularly uses his breaks to catch up with them. Esteban treasures these times of connection, whether unwinding on vacation or spending peaceful time at home.

Another element of Esteban's life that gives him a creative outlet is his passion for music. Despite not being a musician, Esteban loves many different kinds of music and frequently listens to songs when he has free time. He finds inspiration and relaxation in music. Additionally, it helps him relax because racing demands a lot of mental focus, and music offers a welcome mental respite.

Outside of the hectic world of Formula 1, Esteban maintains a well-rounded existence through these interests and pastimes. In addition to keeping him grounded, his ability to mix his professional obligations with extracurricular interests keeps him inspired and career-focused. Esteban's passions—fitness, gaming, travel, and family time—allow him to rejuvenate and discover joy in the small things, which enables him to be the greatest version of himself on and off the track.

Family and Friends

Throughout his Formula 1 career, Esteban Ocon's family and friends have been essential to his success and well-being. From Esteban's early days in karting to his ascent in the motorsport industry, their steadfast emotional and practical support has been essential in helping him overcome the obstacles he has encountered.

Esteban's parents, who have supported him throughout his racing career, are incredibly close to him. Since his early years, his parents, **Sophie** and **Christian**, have provided him with invaluable assistance, guiding him through the intricacies of the racing industry. Esteban has often expressed how much their assistance means to him, particularly during challenging times when he has experienced setbacks or periods of uncertainty. His parents made sacrifices to enable him to enter competitive karting and continue his journey toward Formula 1, allowing him to follow his racing ambitions. Since their efforts were essential in forming Esteban's profession and the man he is now, he freely thanked them.

Additionally, Esteban is very close to his younger sister, **Estelle**. Despite not participating in motorsport, Estelle has consistently been one of Esteban's staunchest fans. The two siblings have a close bond, and Esteban frequently talks about how much he cherishes their time together, mainly when he's not racing. Estelle's presence in Esteban's life serves as a stabilizing influence, reminding him of the value of family as he continues to negotiate the demanding world of Formula 1.

Another crucial component of Esteban's support network is his friends. Esteban has forged strong bonds with several other drivers and members of the motorsport community. Esteban's connection with **Pierre Gasly**, a fellow French driver whom he has known since their early karting days, is one noteworthy example. In addition to being rivals on the track, the two are close friends off it. Their connection goes beyond racing, as the two have supported one another through the highs and lows of their careers and are frequently spotted together at social gatherings. Their friendship serves as a reminder of the value of individuals who are aware of the particular difficulties associated with a career in motorsport and can offer guidance, support, and encouragement to one another.

Over the years, Esteban has also developed enduring bonds with a few of his teammates. Esteban grew close to his teammate **Fernando Alonso** while he was at

Renault and **Alpine**, and the two collaborated to take the team to new heights. They had a competitive relationship, but they also respected each other. Esteban often spoke of his respect for Alonso's background and the knowledge he gained from the two-time world champion. Later, this work-related connection developed, and Esteban praised Alonso as an excellent mentor and inspiration.

Esteban has many acquaintances outside of the racing community who are not active in motorsports. He has often underlined how crucial it is to surround himself with people who enable him to remain normal outside of the spotlight. Despite the demands and constraints of Formula 1, these friendships provide him with a welcome feeling of equilibrium and help him remain grounded.

He tries very hard to spend as much time as possible with his family and friends, even with his hectic travel schedule and time spent on the road. These connections provide him with stability and connectedness, whether via phone conversations, video chats, or in-person visits during race stops. Esteban has frequently attributed his mental toughness to his family and friends, particularly at challenging times, like not having a racing seat in 2019 or overcoming setbacks in his early Formula One career.

In many respects, Esteban's success in Formula 1 results from his skills and the strength of his support network. Every step of the journey has been supported by his family and friends, who have given him guidance, inspiration, and a sense of community in an otherwise lonely world. In both his triumphant and brutal moments, their faith in him has given Esteban the courage to keep going and aim for the sport's top. Esteban's strong bonds with his loved ones constantly remind him of the significance of loyalty, love, and support—fundamental to his identity.

CHAPTER 7: A CHAMPION'S LEGACY

Esteban's Impact on Racing

Beyond his talent and tenacity on the track, Esteban Ocon has significantly influenced racing, especially in Formula 1. His involvement in the sport has affected the culture in the paddock and how racing fans view the sport. Esteban's legacy can be evident in several important areas, including his contribution to team chemistry, his role in promoting French motorsport, and his influence on upcoming generations of drivers.

The fact that Esteban **represents French motorsport** is among the most important facets of his influence. A new generation of young drivers in France has been dramatically influenced by Esteban's success as a French driver in a sport that drivers from other countries have historically dominated. The careers of **Alain Prost** and

René Arnoux were among the high points in French motorsport before Esteban, but his ascent to Formula 1 in the contemporary era rekindle his interest in the sport in his native nation. In addition to upholding the tradition of French drivers, Esteban's ability to continuously compete at a high level and his eventual victory at the 2021 Hungarian Grand Prix have given French drivers hope. His accomplishments demonstrate that anyone can succeed in a demanding sport like Formula 1 with skill, perseverance, and determination, regardless of background.

In addition to his nationality, Esteban has played a significant role in determining the team dynamics of each team he has played for. Esteban has been instrumental in enhancing these teams' performance and competitiveness, whether during his tenure with **Manor**, **Force India**, **Renault**, or **Alpine**. He has been a tremendous benefit because of his versatility, work ethic, and capacity to collaborate with his engineers to develop an automobile. Esteban's attempts to blend in with these groups and help with the car's development have transformed him from a driver into a team player, improving the group's performance. His influence has frequently extended beyond racing to include his ability to function well in a team environment, working with teammates, engineers, and strategists to foster a winning environment.

Esteban's mental toughness and capacity to overcome hardship have significantly influenced how drivers handle obstacles and setbacks. With long hours, high pressure, and the possibility of failure, Formula 1 is a cruel sport. From missing out on a place in 2019 to racing with lower-tier teams, Esteban's career has been challenging. Still, other drivers can learn from his perseverance in overcoming these setbacks. He has demonstrated that perseverance, grit, and the capacity to grow from every encounter can transform obstacles into opportunities. Esteban's experience indicates that in Formula 1, mental and physical toughness are equally as crucial as raw speed.

In addition to being a significant personal accomplishment, his win at the **2021 Hungarian Grand Prix** demonstrated the value of strategy and teamwork in Formula 1. Even though the odds were stacked against him, the race showed how Esteban's bond with his team and his capacity to carry out a racing plan under duress could result in victory. His victory served as the ideal illustration of how a driver can influence a race in ways that go far beyond simply racing quickly; these include making wise choices, staying focused, and cooperating with the team to maximize the car's performance. This triumph demonstrated Esteban's ability to carry out a race plan and showed that individual skill and solid teamwork are frequently necessary for success in Formula 1.

Esteban's beneficial influence on the sport has also been facilitated by his professionalism and conduct off the track. In an environment where scandals and off-track drama frequently eclipse on-track accomplishments, Esteban has managed to uphold a reputation for being dignified, modest, and committed to baseball. He is well-liked by fans and regarded as a respected character in the paddock because of his genuineness and approachability. It is impossible to overstate Esteban's influence as a role model since he keeps demonstrating that ethics, respect for others, and dedication to the sport are just as crucial for success in Formula 1.

Lastly, the legacy he is creating for future generations is another example of Esteban's effect. Esteban inspires future drivers who aspire to compete in Formula 1 because he is a young driver who has established himself in one of the world's most competitive racing settings. His ascent through the motorsport ranks, and his commitment to the sport demonstrate that anyone can achieve the top level of motorsport with perseverance, hard effort, and a love of racing. Aspiring racers can learn from Esteban's experience that there is always room for those who are prepared to strive for their spot, even in a sport that looks unattainable.

Esteban Ocon's driving, demeanor, and attitude have profoundly impacted Formula 1, influencing the sport's future and motivating both drivers and fans. His influence on racing is demonstrated by his capacity to step up, overcome obstacles, and make a difference on and off the track. With every race, Esteban continues to influence the direction of racing, whether as a competitor, teammate, or role model.

The Future of Racing

Esteban Ocon's contribution to influencing the direction of racing is becoming increasingly important as his Formula 1 career develops. Racing, particularly in Formula 1, has an exciting future filled with innovation, sustainability, and heightened competition. Beyond only competing on the track, Esteban is involved in this future through his contributions to the sport's development, his work on the creation of new technologies, and his influence on the next generation of drivers.

The transition to sustainability is one of the most significant adjustments that racing will undergo. By introducing **hybrid engines** and aiming to be **carbon-neutral by 2030**, Formula 1 is already trying to lessen its environmental effect. Along with other drivers, Esteban Ocon is increasingly participating in the sport's initiatives to innovate and create technology that helps create a more sustainable future. Esteban's involvement in this shift will be essential as the sector concentrates on alternative fuels and hybrid technology. His curiosity about how technology may enhance a car's performance while reducing its environmental impact aligns with Formula 1's larger goal of making motorsport more environmentally friendly.

A key component of racing's future is the technological developments in race vehicle design**. Modern technology makes Formula 1 a sport, and drivers like Esteban are instrumental in creating new materials and systems that will shape the sport's future. Future automobiles will have better aerodynamics and more intelligent data analytics, making them safer, quicker, and more effective than ever. Esteban's input is invaluable in improving these technological advancements, particularly concerning the car's setup and performance. Esteban's background in car development will continue to advance the sport as Formula 1 strives for ever-increasing accuracy and economy.

The increased competition in Formula 1 is another factor that will influence racing in the future, in addition to technological developments. The level of competition will increase as more teams and skilled drivers join the sport, making races even more thrilling and unexpected. Esteban's career exemplifies the competitive mentality that will shape racing's future because of his unwavering resolve to compete for every position and seize every chance. The sport is about to enter a period of increased intensity as more youthful, dynamic drivers join Formula 1 every year. Esteban's capacity to adjust and develop will be crucial to staying up with the growing competition.

The future of racing will also be shaped by creating new race formats and creative fan interaction techniques. With events like the Sprint Races and the possibility of more engaging spectator experiences, Formula 1 is already looking into new strategies to appeal to a larger audience. Esteban will play a significant role in these adjustments as a well-known character in the sport. The racing community will continue to be involved in fresh and exciting ways because of his social media presence, dedication to fan interaction, and role as a representative of the sport. Esteban's contribution to bridging the divide between the sport's elite world and its ardent fan base will grow significantly as it becomes more widely available.

The use of **artificial intelligence (AI)** and **machine learning** in race strategy is another element influencing the direction of racing. AI is already being used to evaluate race data and forecast results, providing drivers and teams with a more accurate picture of their performance during competitions. Esteban is well-suited for this future when data-driven tactics will be even more critical to a race's success, thanks to his keen decision-making abilities and capacity for tight collaboration with his engineers. Esteban's flexibility and readiness to adopt new technology will be essential in negotiating this shift as artificial intelligence plays an increasingly more significant role in racing.

The importance of mental and physical preparation will also keep changing in the future of racing. To thrive in Formula 1, drivers will need optimal physical and psychological health as the demands of the sport rise. Esteban's dedication to maintaining his physical and emotional health will be a model for following drivers who must prepare more thoroughly. Future drivers may need to learn mindfulness, cognitive performance, stress management techniques, and traditional fitness regimens to cope with the sport's mounting demands. Esteban is already laying the foundation for the future of racing preparation with his commitment to maintaining his physical and mental toughness.

Finally, there will be more international cooperation and more women in motorsport in the future of racing. In the years to come, there will probably be greater diversity in Formula 1, which is already growing more inclusive. As part of this change, Esteban, who has always respected his colleagues and rivals, will work with and encourage various newcomers to the sport. His efforts to foster inclusivity and create a more hospitable racing atmosphere will influence the sport's culture for upcoming generations.

Beyond his career, Esteban Ocon significantly influences the direction of racing. Esteban will continue to shape the sport's future with his commitment to sustainability, technology, competition, fan interaction, and personal development. Esteban will continue to play a significant role in the sport's exciting future as he manages his career and adjusts to the shifting conditions of Formula 1, motivating teams, drivers, and fans for many years to come. With drivers like Esteban at the forefront, bringing new ideas, vitality, and an unwavering will to succeed, the future of racing appears brighter.

CONCLUSION

Esteban Ocon's experience in Formula 1 has been one of tenacity, willpower, and enthusiasm. Esteban's journey from his modest karting beginnings to his ascent to fame as a Formula 1 driver proves the value of never giving up on your goals and the strength of tenacity. In addition to showcasing his extraordinary talent on the track, he has seized the chance to encourage, inspire, and represent the upcoming generation of racers.

His influence goes well beyond his racing accomplishments. Through his dedication to sustainability, contributions to automobile development, and impact on racing culture, Esteban has shaped the future of motorsport and solidified his status as a driver who will continue to impact the sport for years to come. He sets an example for spectators and aspiring drivers by embodying sportsmanship, humility, and collaboration.

Esteban's capacity for adaptation, learning, and development guarantees that his legacy will live on as the sport changes to accommodate new opportunities, difficulties, and technological advancements. Esteban Ocon is more than just a Formula 1 driver; he is a role model representing what is possible with perseverance, hard work, and a passion for the sport. This is evident in his ambition to succeed, his commitment to developing as a driver, and his focus on positively influencing both on and off the track.

Esteban has a long way to go in his adventure. With every race, season, and challenge, he keeps pushing the limits of what is feasible in Formula 1. Esteban Ocon will surely be recognized as one of the drivers who contributed to the development of the modern era of motorsport as he forged ahead for upcoming generations of drivers.

www.ingramcontent.com/pod-product-compliance
Lightning Source LLC
Chambersburg PA
CBHW071654240526
45469CB00023B/2373